Lab Manual – Universal Fluid Power Trainer
HSV2/MSOE02

Electro-Hydraulic Components and Systems

MS OE

UNIVERSITY

www.msoe.edu/seminars

Lab Manual
Universal Fluid Power Trainer (UFPT)
Electro-Hydraulic Components and Systems

ISBN: 978-0-9977816-8-7

Copyright © 2016 by
Milwaukee School of Engineering

Printed in the United States of America

Lab Manual – Universal Fluid Power Trainer
Electro-Hydraulic Components and Systems

Lab Manual-UFPT

How to Use Me Safely?

1- Emergency Stop:
- Locate the emergency stop.
- Press the emergency stop immediately if you feel any dangerous situation e.g. leakage, bad smell, system instability, unusual noise

2- Warning Light: consult the instructor if any of red led lights is turned ON.

3- Consult the Instructor: if you are not sure about any thing.

Please, no use of flash drives

Safety Regulations, Contd.

4. Safety Glasses: Wear the safety glass and shoes where applicable.

5. Hydraulic Hoses: Use **proper length &** respect minimum **bend radius.**

6. Quick Disconnect: Make sure it is perfectly connected **"hear the click".**

7. Trapped pressure: Connect high pressure measurement hose to the tank first, then to the point at which the pressure is trapped.

8. Accumulator (stored energy hazard): Isolate the accumulator whenever it is not needed.

9. Pressure Measurement Gauges: P4 assigned for tank pressure only using female-female hose between the tank header and P4.

10. Drain: make sure drain lines are plumbed where needed.

11. Flow-meter: is a unidirectional flow meter, follow the arrow.

12. Turning ON any power : Warn all team members first.

13. Circuit (Hydraulic/pneumatic/24V) Modification: turn the corresponding power button Off first.

Safety Regulations, Contd. For courses other than Introduction

---

14. Cables & Sockets: Keep the unused electrical sockets covered.

15. Cables & Sockets: Do not stretch or pull from wires & make sure it is correctly fit.

16. Servo Valve Cable: Servo valve cable contains DC/DC convertor.

17. Banana Jacks Cables: Use proper length of cables.

18. Electrical 24V Circuits: In order not to get confused and to avoid a short circuit, the good advice is to connect one vertical line at a time.

19. Pressure Sensors: make sure it is tightened enough to sense the pressure. High pressure sensors are numbered (1/4 – 3/4) and one low pressure is tagged (1/4).

20. Pneumatic Power: turn air valves off before disassemble air hose.

21. Compressor(stored energy hazard): Before turning the compressor ON, make sure air valves are off.

22. Follow up the "lab procedure".

23. Follow the machine "Machine Startup Procedure".

24. Follow the machine "Machine Shutdown Procedure".

Lab Procedure

1. Form work groups. A group works together till the end of the class..
2. Listen to the instructor orientation. Ask questions if needed.
3. Read the lab instructions line by line even after the orientation.
4. Read schematics (Electrical/Hydraulics) and work instructions.
5. Build the system as per the step-by-step given instructions.
6. Operate the circuits and follow the given instructions.
7. Recorded your observations as per the exercise instructions.
8. Analyze the data and answer the posted questions.
9. Disconnect the circuit and put down hoses and measuring instruments at the end of every lab session.
10. Store back the components as per its drawer identification.

Lab Procedure, Contd.

Hydraulic Power Supply

Adjustment of PRV at the beginning of every exercise

Matlab RT Models

1-Activating, 2-Connecting 3-Running

Matlab

Current Folder: C:\11-Common Matlab\Models Real Time

Simulink   100   External

Machine Startup Procedure

1. Casters: Lock the casters to prevent accidental move of the unit.
2. Uncoil the power cords and plug the unit to the electrical wall outlet.
3. Take the hoses out of the hose bin and hang them on the side hooks.
4. Turn ON the electrical control panel.
5. Turn ON the HMI (only when needed).
6. Turn ON the printer (only when needed).

6

Machine Shutdown Procedure

1. Make sure cylinders are fully retracted.
2. Discharge the accumulator from the manual discharge valve.
3. Discharge the compressor by opening one of the air valves.
4. Turn OFF the pump, air power, and control power.
5. Shutdown the computer from the windows.
6. Turn OFF the mouse and store it back in drawer "A"
7. Turn OFF the 24 Volt power supply.
8. Turn OFF the printer.
9. Set toggle switches to 'Pot" position.
10. Set the potentiometers to "min".
11. Unplug the power cord and hang it under the counter.
12. Unplug any remaining electrical connections.

7

Machine Shutdown Procedure, Contd.

13. Disconnect hydraulic hoses, cover their ends by dust caps, and bring the store hoses back to the hose bin.
14. Cover all hydraulic components with dust caps.
15. Store the components as per its drawer identification.
16. Please, clean the safety glasses and store them.
17. Please, clean spilled oil.

Exit

8

MSOE – Fluid Power and Motion Control Professional Education

Lab20: Identity (YES Function) by Indirect Activation

Objective:

This lab is to practice building an electro-hydraulic circuit to drive a cylinder based on a push button pressing.

Step 1: Prepare Components:

MSOE#-X	Component Name	QTY	Drawer #
N027	Pressure measurement hose	1	B
N030	4/2 solenoid operated DCV	1	D3
N035	Solenoid cables with LED	1	D3
N054	Banana jack cables		NA

Step 2: Adjust the PRV to 500 psi:

1. Connect the pressure measurement hose as shown to P1.
2. Set the Pp and Qp toggle switches to "Pot" positions.
3. Adjust pump potentiometers, Pp and Qp, to maximum positions.
4. Open the PRV halfway.
5. Turn ON the hydraulic power.
6. Close the PRV gradually until the pressure gauge reads the set value.
7. Turn OFF the hydraulic power.

MSOE – Fluid Power and Motion Control Professional Education

Step 3: Electrical Circuit:

1. Make sure the 24 Volt s Off.
2. Build the shown electrical circuit.
3. Turn the 24 Volt ON.
4. Check proper functioning of the electrical circuit by observing the solenoid cable's led lighten when the PB is pressed.
5. If it works fine, connect the solenoid cable to the valve.

Step 4: Hydraulic Circuit:

1. Build the shown hydraulic circuit.
2. Turn the hydraulic power ON.
3. Practice extending the cylinder by pressing and releasing the PB.
4. Share your observation with the instructor.
5. Turn the hydraulic power OFF

You may repeat the exercise by the use of toggle switch instead of the bush button

MSOE – Fluid Power and Motion Control Professional Education

Lab21: Signal Storage by Electrical Latching

Objective:

A cylinder is to be extended by means of pressing a push-button. When the push-button is released the cylinder is to continue to extend by means of signal storage until the end position is reached. The retraction of the cylinder may only be possible when the signal storage is reset by means of a second push-button.

Step 1: Prepare Components:

MSOE#-X	Component Name	QTY	Drawer #
N027	Pressure measurement hose	1	B
N030	4/2 solenoid operated DCV	1	D3
N035	Solenoid cables with LED	1	D3
N054	Banana jack cables		NA

0

Step 2: Adjust the PRV to 500 psi:

1. Connect the pressure measurement hose as shown to P1.
2. Set the Pp and Qp toggle switches to "Pot" positions.
3. Adjust pump potentiometers, Pp and Qp, to maximum positions.
4. Open the PRV halfway.
5. Turn ON the hydraulic power.
6. Close the PRV gradually until the pressure gauge reads the set value.
7. Turn OFF the hydraulic power.

1

MSOE – Fluid Power and Motion Control Professional Education

Step 3: Electrical Circuit:

1. Make sure the 24 Volt s Off

2. Build the shown electrical circuit.

3. Turn the 24 Volt ON

4. Check proper functioning of the electrical circuit by observing the solenoid cable's led lighten when the PB-S1 is pressed and turned off when the PB-S2 is pressed

5. If it works fine, connect the solenoid cable to the valve.

2

3

MSOE – Fluid Power and Motion Control Professional Education

Step 4: Hydraulic Circuit:

1. Build the shown hydraulic circuit.
2. Turn the hydraulic power ON.
3. Practice extending and retracting the cylinder.
4. Share your observation with the instructor.
5. Turn the hydraulic power OFF.

Lab22: Electrical Protection of Two Solenoids Valve

Exit

Objective:

A hydraulic motor is driven in both directions by three-position two-solenoids directional valve. Once one of the PBs is pressed, pressing the other PB will do nothing. A "Stop" PB must be pressed first before change the direction of motor rotation.

Step 1: Prepare Components:

MSOE#-X	Component Name	QTY	Drawer #
N027	Pressure measurement hose	1	B
N031	4/3 tandem center solenoid operated DCV	1	D3
N035	Solenoid cables with LED	2	D3
N054	Banana jack cables		NA

0

Step 2: Adjust the PRV to 500 psi:

1. Connect the pressure measurement hose as shown to P1.
2. Set the Pp and Qp toggle switches to "Pot" positions.
3. Adjust pump potentiometers, Pp and Qp, to maximum positions.
4. Open the PRV halfway.
5. Turn ON the hydraulic power.
6. Close the PRV gradually until the pressure gauge reads the set value.
7. Turn OFF the hydraulic power.

1

MSOE – Fluid Power and Motion Control Professional Education

Step 3: Electrical Circuit:

1. Make sure the 24 Volt s Off.

2. Build the shown electrical circuit.

3. Turn the 24 Volt ON.

4. Check proper functioning of the electrical circuit by observing the solenoid cable's led lighten when the corresponding push button is pressed

5. If it works fine, connect the solenoid cables to the valve.

MSOE – Fluid Power and Motion Control Professional Education

Step 4: Hydraulic Circuit:

1. Build the shown hydraulic circuit.
2. Turn the hydraulic power ON.
3. Practice driving the motor in both directions by pressing the two push buttons (one at a time)
4. Share your observation with the instructor
5. Turn the hydraulic power OFF

MSOE – Fluid Power and Motion Control Professional Education

Lab23: Position -Dependent Cylinder Deceleration

Objective:

The horizontal cylinder extends by pressing S1. The cylinder extends with maximum speed and decelerates when the position switch is activated. The circuit is reset by pressing S3 first. Then the cylinder is retracted at max speed for full stroke by pressing S2.

Step 1: Prepare Components:

MSOE#-X	Component Name	QTY	Drawer #
N027	Pressure measurement hose	1	B
N030	4/2 solenoid operated DCV	1	D3
N038	4/3 closed center solenoid operated DCV	1	D3
N035	Solenoid cables with LED	3	D3
N007	Throttle-Check Valve	1	D2
N028	Inductive Proximity Switches	1	D1
N054	Banana jack cables		NA

Step 2: Adjust the PRV to 500 psi:

1. Connect the pressure measurement hose as shown to P1.
2. Set the Pp and Qp toggle switches to "Pot" positions.
3. Adjust pump potentiometers, Pp and Qp, to maximum positions.
4. Open the PRV halfway.
5. Turn ON the hydraulic power.
6. Close the PRV gradually until the pressure gauge reads the set value.
7. Turn OFF the hydraulic power.

Step 3: Electrical Circuit:

1. Make sure the 24 V is Off.

2. Build the shown electrical circuit.

3. Shift the right proximity switch to position of 5 inch stroke.

4. Turn the 24 V ON.

5. Check proper functioning of the electrical circuit by observing the solenoid cable's led lighten when the corresponding push button is pressed.

6. Check the proper functioning of the proximity switch by observing the corresponding solenoid lightening when the proximity switch is activated.

7. If the electrical circuit is working properly, then connect the solenoid cables to the valves.

Note 2: Y1 is the solenoid beside the parallel position of the valve.

2

3

Step 4: Hydraulic Circuit:

1. Build the shown hydraulic circuit.

2. Turn the hydraulic power ON.

3. Practice driving the cylinder in both directions by pressing the two push buttons S1 and S2 (one at a time and after reset by S3).

4. Adjust the cylinder deceleration rate by cranking the throttle valve.

5. Share your observation with the instructor.

6. Turn the hydraulic power OFF.

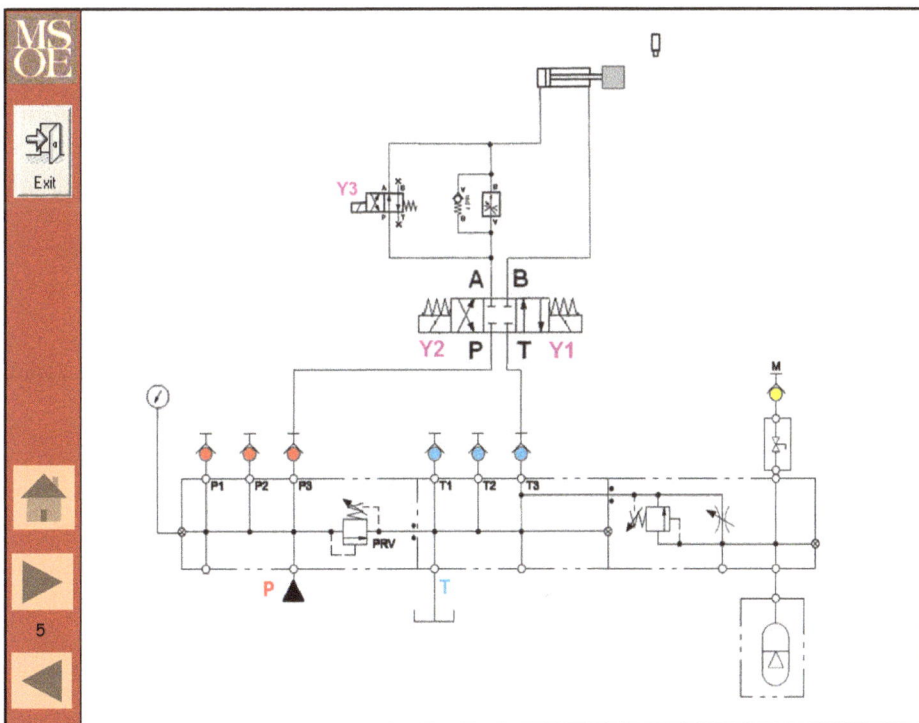

MSOE – Fluid Power and Motion Control Professional Education

Lab24: One-Cycle Hydraulic Cylinder Reciprocation

Objective:

This lab is to practice Pressure-Dependent Cylinder Reversal as an application of one-cycle cylinder reciprocation. In this experiment, a cylinder is to extend upon the operation of a push-button. The cylinder retracts automatically when a pressure switch is activated.

Step 1: Prepare Components:

MSOE#-X	Component Name	QTY	Drawer #
N027	Pressure measurement hose	1	B
N030	4/2 solenoid operated DCV	1	D3
N035	Solenoid cables with LED	1	D3
N032	Pressure Switch + cable (banana ends)	1	D1
N054	Banana jack cables		NA

Step 2: Adjust the PRV to 500 psi:

1. Connect the pressure measurement hose as shown to P1.
2. Set the Pp and Qp toggle switches to "Pot" positions.
3. Adjust pump potentiometers, Pp and Qp, to maximum positions.
4. Open the PRV halfway.
5. Turn ON the hydraulic power.
6. Close the PRV gradually until the pressure gauge reads the set value.
7. Turn OFF the hydraulic power.

Step 3: Electrical Circuit:

1. Make sure the 24 V is OFF.

2. Build the shown electrical circuit.

3. Connect the solenoid cable to the solenoid valve.

4. Turn 24 V ON.

5. Check the proper functioning of activating the solenoid by pressing S1 and deactivating it by pressing S2.

6. If it works fine, connect the solenoid cable to the valve.

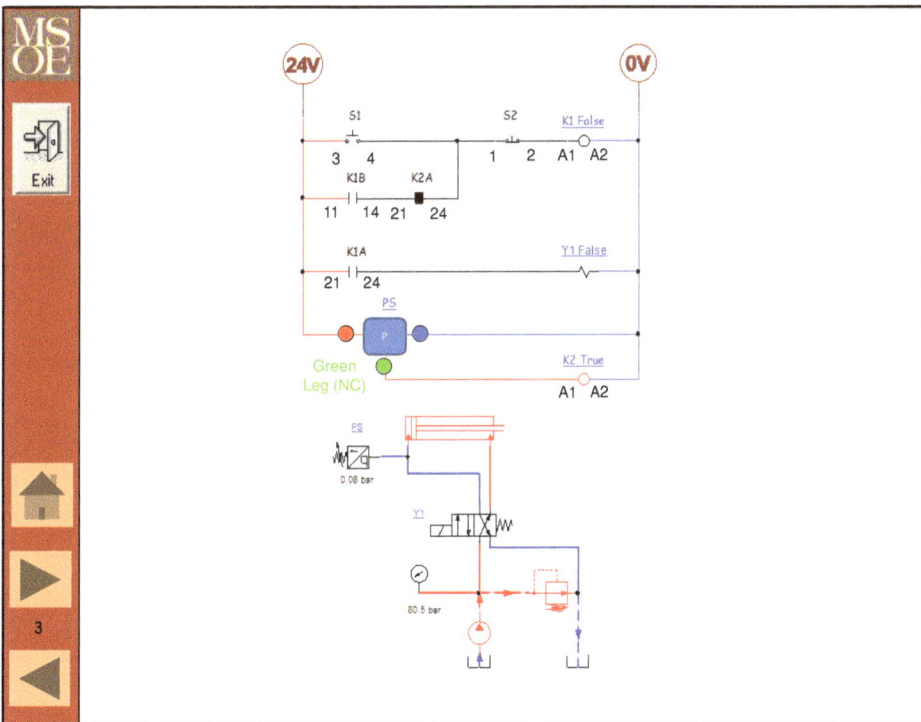

MSOE – Fluid Power and Motion Control Professional Education

Step 4: Hydraulic Circuit:

1. Build the shown hydraulic circuit.

2. Turn the hydraulic power ON.

3. Practice moving the cylinder by pressing the push buttons, cylinder should retract once it is deadheaded and pressure reaches 500 psi.

4. Share your observation with the instructor

5. Turn the hydraulic power OFF

Note:
- Make sure Pressure Switch adjusted as follows:
- Out 1 (Black) = NO, SP1 = 500 psi, RP1 = 250.
- Out 2 (Green) = NC, SP1 = 500 psi, RP1 = 250.
- Adjustments instructions are in the manual page 8/1008.

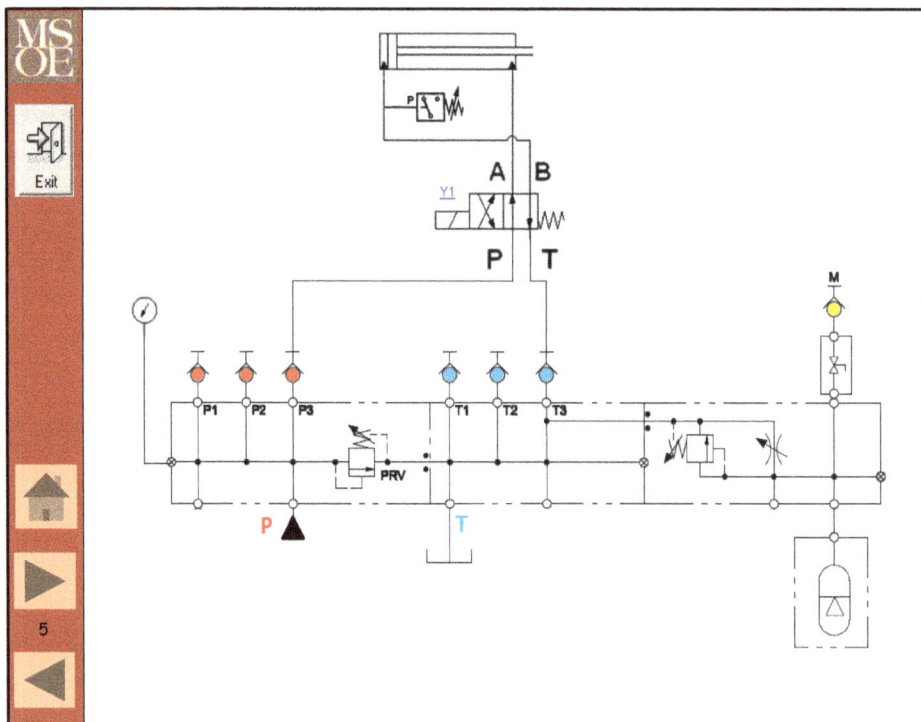

MSOE – Fluid Power and Motion Control Professional Education

Lab25: Panic Circuit

Exit

Objective:

In this experiment, a warning red light and panic sound will be energized when the vertical cylinder is lifted. Green light will be ON if the cylinder retracted.

Step 1: Prepare Components:

MSOE#-X	Component Name	QTY	Drawer #
N027	Pressure measurement hose	1	B
N030	4/2 solenoid operated DCV	1	D3
N035	Solenoid cables with LED	1	D3
N054	Banana jack cables		NA

0

Step 2: Adjust the PRV to 500 psi:

1. Connect the pressure measurement hose as shown to P1.
2. Set the Pp and Qp toggle switches to "Pot" positions.
3. Adjust pump potentiometers, Pp and Qp, to maximum positions.
4. Open the PRV halfway.
5. Turn ON the hydraulic power.
6. Close the PRV gradually until the pressure gauge reads the set value.
7. Turn OFF the hydraulic power.

Exit

1

Step 3: Electrical Circuit

1. Make sure the 24 V is OFF
2. Build the shown electrical circuit
3. Connect the solenoid cable (Banana Ends) to the solenoid valve.
4. Turn 24 V ON
5. Check the proper functioning of activating/deactivating the panic sound and light by pressing and releasing S1
6. If it works fine, connect the solenoid cable to the valve

Step 4: Hydraulic Circuit:

1. Build the shown hydraulic circuit.
2. Turn the hydraulic power ON.
3. Practice extending the cylinder by pressing the push buttons S1 and the cylinder retracts when the PB is released.
4. Share your observation with the instructor.
5. Turn 24 V Off.

Lab26: Cylinder Motion Control Performance using Switching Valve versus Proportional Valve

Exit

Objective:

This lab is to demonstrate the enhanced performance of a position controlled hydraulic cylinder when a proportional/servo valve is used versus On/Off valve.

Step 1: Prepare Components:

MSOE#-X	Component Name	QTY	Drawer #
N027	Pressure measurement hose	1	B
N029	Proportional DCV + Cable	1	D5
N030	4/2 solenoid operated DCV	1	D3
N035	Solenoid cables (M12)	1	D3
N042 (1-3)/4	Pressure transducer	1	D1
N053	BNC Cable	1	D1

0

Step 2: Adjust the PRV to 500 psi:

1. Connect the pressure measurement hose as shown to P1.
2. Set the Pp and Qp toggle switches to "Pot" positions.
3. Adjust pump potentiometers, Pp and Qp, to maximum positions.
4. Open the PRV halfway.
5. Turn ON the hydraulic power.
6. Close the PRV gradually until the pressure gauge reads the set value.
7. Turn OFF the hydraulic power.

1

Step 3: Electrical Circuit:

1. Connect solenoid cable to DO6.
2. Connect pressure sensor socket to AIO8.

Step 4: Hydraulic Circuit:

1. Build the shown hydraulic circuit.
2. Connect pressure sensor to valve port P
3. Turn the hydraulic power ON.

Lab Animation: C044

MSOE – Fluid Power and Motion Control Professional Education

Step 5: Data Acquisition:

1. Activate, connect RT and run RTModel014.mdl

2. Model will stop by itself after 15 s.

3. Turn OFF the hydraulic power.

4. Share your observation with the instructor. Notice the pressure spikes.

Step 6: Electrical Circuit:

1. Connect the proportional valve to the socket AI14V1 and the other side of the cable to the valve.
2. Connect the BNC cable as shown below.
3. Put the toggle switch of V1 at "Ext." position.

Step 7: Hydraulic Circuit:

1. Build the shown hydraulic circuit.
2. Relocate the pressure sensor to the proportional valve P port.
3. Turn the hydraulic power ON.
4. Cylinder must be fully retracted. If not, put V1 toggle switch at "Pot." position, move the potentiometer to retract the cylinder then return back the toggle switch to "Ext." position.

Lab Animation: C045

Step 8: Data Acquisition:

1. Activate, connect RT and run RTModel015.mdl

2. Model will automatically drive the cylinder and stop after 15 s.

3. Turn OFF the hydraulic power.

4. Share your observation with the instructor.

Lab summery

1- Using proportional/servo valve improve the cylinder motion profile

☐ TRUE   ☐ FALSE

2- Using proportional/servo valve reduce the pressure spikes

☐ TRUE   ☐ FALSE

Exit

8

Lab27: Cylinder Motion Control Performance using Servo Valve versus Proportional Valve

Exit

Objective:
This lab is to observe the difference in cylinder motion control performance using servo valve versus proportional valve.

Step 1: Prepare Components:

MSOE#-X	Component Name	QTY	Drawer #
N027	Pressure measurement hose	1	B
N029	Proportional DCV + Cable	1	D5
N033	Servo Valve + Servo Valve Cable	1	D5
N053	BNC Cable	2	D1

0

Step 2: Adjust the PRV to 500 psi:

Exit

1. Connect the pressure measurement hose as shown to P1.
2. Set the Pp and Qp toggle switches to "Pot" positions.
3. Adjust pump potentiometers, Pp and Qp, to maximum positions.
4. Open the PRV halfway.
5. Turn ON the hydraulic power.
6. Close the PRV gradually until the pressure gauge reads the set value.
7. Turn OFF the hydraulic power.

1

MSOE – Fluid Power and Motion Control Professional Education

Step 3: Electrical Circuit:

1. Connect the proportional valve to the socket AI14V1 and the other side of the cable to the valve.
2. Connect the servo valve to the socket AI15V2 and the other side of the cable to the valve.
3. Connect the BNC cables as shown below.
4. Put the toggle switches of V1 and V2 at "Ext." positions.

AI14 V1

AI15 V2

Step 4: Hydraulic Circuit-Use the Servo Valve

1. Build the shown hydraulic circuit.
2. Turn the hydraulic power ON.
3. Cylinder must be fully retracted. If not, put V2 toggle switch at "Pot." position, move the potentiometer to retract the cylinder then return back the toggle switch to "Ext." position.

MSOE – Fluid Power and Motion Control Professional Education

Step 5: Data Acquisition:

1. Activate, connect RT and run RTModel021.mdl

2. Model will automatically drive the cylinder and stop after 20 s.

3. Share your observation with the instructor.

4. Turn OFF the hydraulic power.

Step 6: Hydraulic Circuit-Use the Proportional Valve

1. Build the shown hydraulic circuit.
2. Turn the hydraulic power ON.
3. Cylinder must be fully retracted. If not, put V1 toggle switch at "Pot." position, move the potentiometer to retract the cylinder then return back the toggle switch to "Ext." position.

Note: pay attention to port connection

Lab Animation: C045

Step 7: Data Acquisition:

1. Activate, connect RT and run RTModel022.mdl

2. Model will automatically drive the cylinder and stop after 20 s.

3. Turn OFF the hydraulic power.

4. Share your observation with the instructor.

Servo Valve

Proportional Valve

Lab summery

1- Which of the following valves show zero SSE?

☐   Proportional Valve   ☐   Servo Valve

Why?

Exit

8

MSOE – Fluid Power and Motion Control Professional Education

Lab28: Digital Control of Variable Displacement Pumps

Objective:
This lab is to demonstrate the digital control of an EH-controlled variable displacement pumps.

Step 1: Prepare Components:

MSOE#-X	Component Name	QTY	Drawer #
N027	Pressure measurement hose	1	B
N016-TX-2/4	Flowmeter (Medium Scale)	1	D2
N036	Proportional FCV + Cable	1	D5
N053	BNC Cable	2	D1

Exit

0

Step 2: Adjust the PRV to 500 psi:

1. Connect the pressure measurement hose as shown to P1.
2. Set the Pp and Qp toggle switches to "Pot" positions.
3. Adjust pump potentiometers, Pp and Qp, to maximum positions.
4. Open the PRV halfway.
5. Turn ON the hydraulic power.
6. Close the PRV gradually until the pressure gauge reads the set value.
7. Turn OFF the hydraulic power.

Exit

1

MSOE – Fluid Power and Motion Control Professional Education

Step 3: Electrical Circuit

Case A: Practice Pressure Compensation Digital Control

1. Connect the proportional flow control valve cable to socket AI14V1 and the other side of the cable to the valve.
2. Put the toggle switch of Qmax at "Pot/Max" position.
3. Put the toggle switch of Pmax at "Ext." position.
4. Put the toggle switch of V1 at "Ext." position.
5. Connect the BNC Cables as shown below.

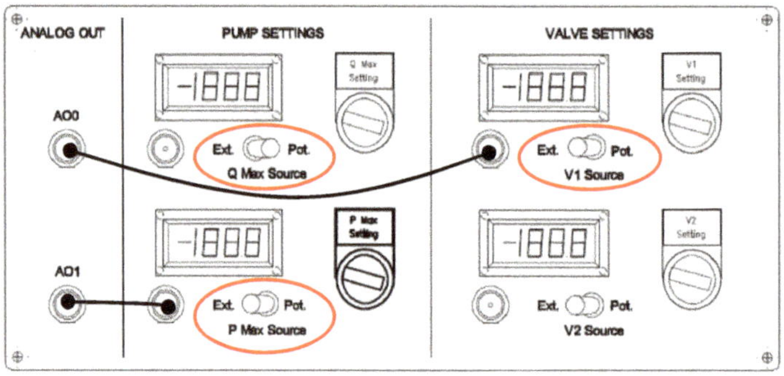

Step 4: Hydraulic Circuit:

1- Build the shown hydraulic circuit.
- Note: PFCV is 2-way mode (i.e. port p is disconnected).
2- Turn the hydraulic power ON.

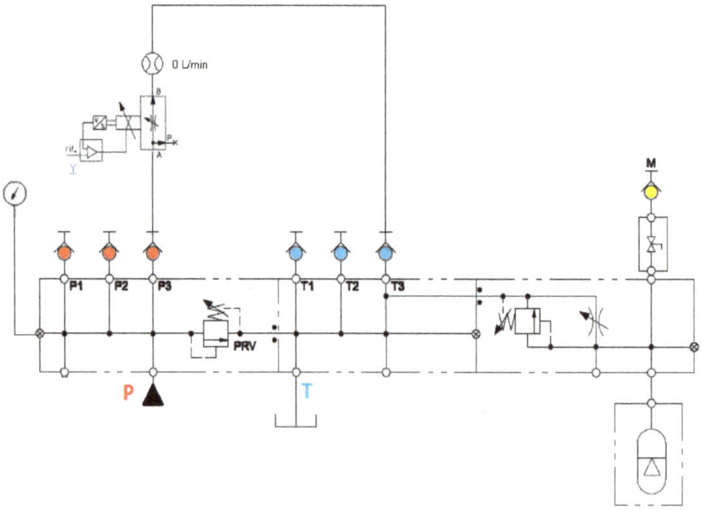

Step 5: Data Acquisition,

1. Activate Matlab Model: RTModel012A.mdl

2. Make sure to set the digital pressure compensator to 400 psi.

3. Connect the Matlab RT and run the model.

4. Model will automatically draw the pump flow against variable load and stop after 100 s.

5. Share your observation with the instructor.

6. Repeat the process at new P_{max} = 300 (psi).

7. Share your observation with the instructor.

8. Turn OFF the hydraulic power.

4

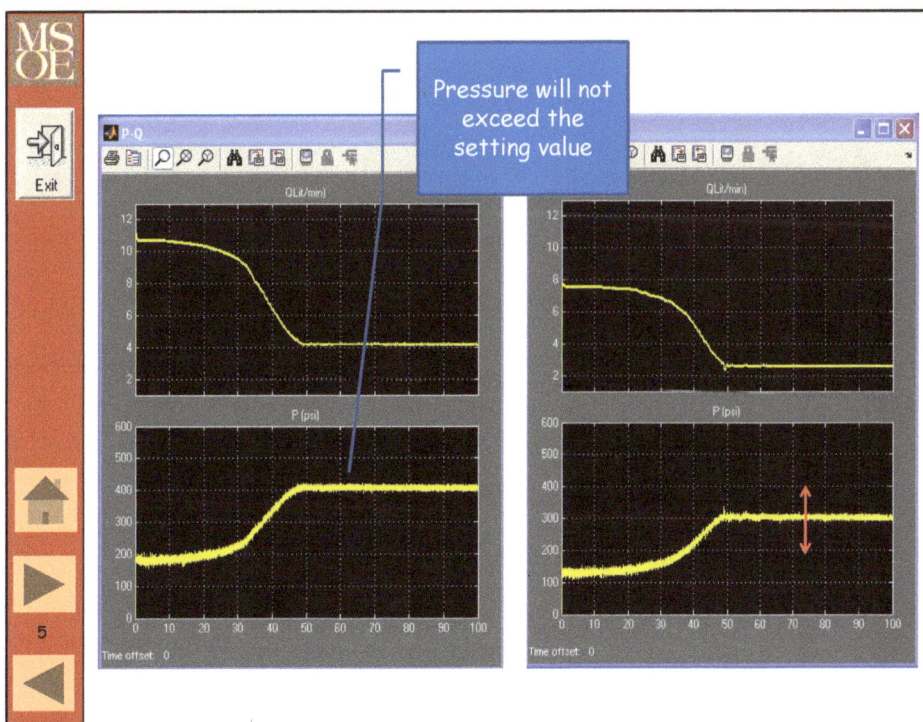

Pressure will not exceed the setting value

Step 6: Electrical Circuit

Case B: Practice Pump Flow Digital Control

1. Put the toggle switch of Qmax at "Ext." position.
2. Put the toggle switch of Pmax at "Pot/Max" position.
3. Keep the toggle switch of V1 at "Ext." position.
4. Connect the BNC Cables as shown below.

Step 7: Data Acquisition,

1. Activate Matlab Model: RTModel012B.mdl

2. Make sure Qmax = 6 (Lit/min).

3. Connect Matlab RT and run the model.

4. Model will automatically draw the pump flow against variable load and stop after 100 s.

5. Share your observation with the instructor.

6. Repeat the process at new Qmax = 4 (Lit/min).

7. Share your observation with the instructor.

MSOE – Fluid Power and Motion Control Professional Education

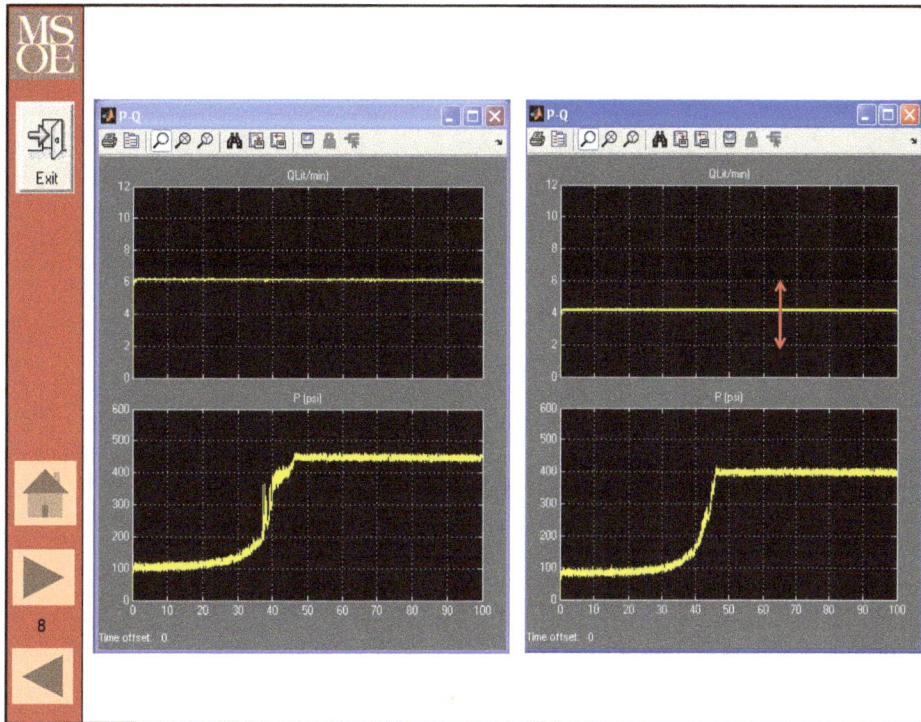

Step 8: Electrical Circuit

Case C: Practice Pump Power Limiting Digital Control

Keep same connection and settings

1. Put the toggle switch of Qmax at "Ext." position.
2. Put the toggle switch of Pmax at "Pot/Max" position.
3. Keep the toggle switch of V1 at "Ext." position.
4. Connect the BNC Cables as shown below.

MSOE – Fluid Power and Motion Control Professional Education

Step 9 : Data Acquisition, Power Limiting

1. Activate Matlab Model: RTModel012C.mdl

2. Make sure HPmax = 0.5 (HP)

3. Connect and run the Matlab RT

4. Model will automatically draw the pump flow against variable load and stop after 100 s.

5. Share your observation with the instructor.

6. Repeat the process at new HPmax = 0.25 (HP)

7. Share your observation with the instructor.

8. Turn OFF the hydraulic power

10

Check the power
= (P × Q × 1.36)/(14.5 × 600)
= (410 × 7 × 1.36) / (14.5 × 600)
= 0.51 HP

Check the power
= (P × Q × 1.36)/(14.5 × 600)
= (280 × 5.8 × 1.36) / (14.5 × 600)
= 0.25 HP

11

Lab29: Digital Control of a Hydraulic Cylinder Position

Objective:
This lab is to practice building digital control functions for a hydraulic cylinder position

Step 1: Prepare Components:

MSOE#-X	Component Name	QTY	Drawer #
N027	Pressure measurement hose	1	B
N029	Proportional DCV + Cable	1	5
N053	BNC Cable	1	1

Exit

0

Step 2: Adjust the PRV to 500 psi:

1. Connect the pressure measurement hose as shown to P1.
2. Set the Pp and Qp toggle switches to "Pot" positions.
3. Adjust pump potentiometers, Pp and Qp, to maximum positions.
4. Open the PRV halfway.
5. Turn ON the hydraulic power.
6. Close the PRV gradually until the pressure gauge reads the set value.
7. Turn OFF the hydraulic power.

Exit

1

Step 3: Electrical Circuit:

AI14 V1

1. Connect the proportional valve to the socket AI14V1 and the other side of the cable to the valve.
2. Connect the BNC cable as shown below
3. Put the toggle switch of V1 at "Ext." position

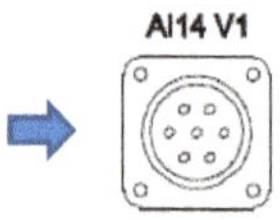

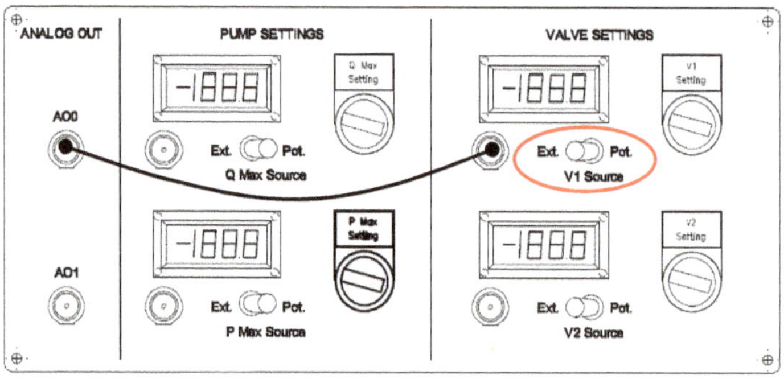

Step 4: Hydraulic Circuit:

1. Build the shown hydraulic circuit
2. Turn the hydraulic power ON.
3. Cylinder must be fully retracted. If not, put V1 toggle switch at "Pot." position, move the potentiometer to retract the cylinder then return back the toggle switch to "Ext." position

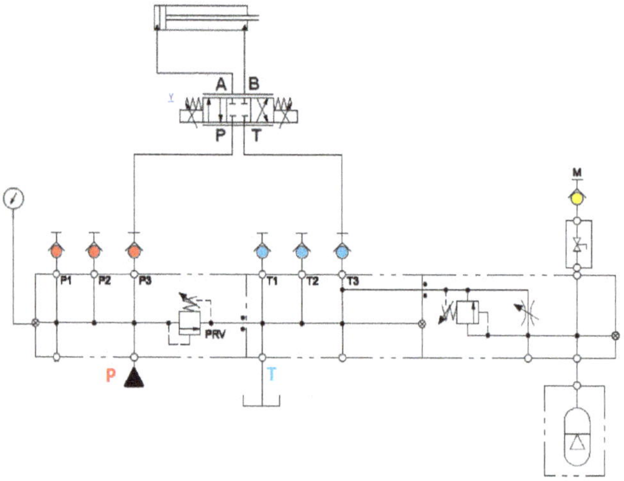

Step 5: Data Acquisition: (Proportional Gain Effect)

1. Activate the Model: RTModel013.mdl

2. Click on the "Amplifier" model and set the proportional gain to (1). Make sure the signal bypasses the dead band eliminator and ramp generator.

3. Cylinder must be fully retracted. If not, put V1 toggle switch at "Pot." position, move the potentiometer to retract the cylinder then return back the toggle switch to "Ext." position.

4. Connect the Matlab RT and run the model.

5. The model will drive the cylinder and stops after 10 second. Record the steady state error and the settling time.

6. Switch the toggle switch to "Pot." position and use the potentiometer to retract the cylinder, then bring the toggle switch back to "Ext."

7. Repeat the steps from 1 through 6 with an increment increase of the proportional gain of 1 up to a gain of 5.

8. Share your observation with the instructor and answer the question.

❏ Settling Time

Gain/Group	G1	G2	G3	G4
1				
2				
3				
4				
5				

❏ Steady State Error

Gain/Group	G1	G2	G3	G4
1				
2				
3				
4				
5				

❑ Increasing the proportional gain increases the speed of the system response to a step input ?

☐ TRUE   ☐ FALSE

❑ Increasing the proportional gain reduces the settling time?

☐ TRUE   ☐ FALSE

6

Step 6: Data Acquisition: (Valve Dead Band Eliminator)

First we need to find the dead band value of the valve

❑ Set the toggle switch of the proportional valve to "Pot." position.

❑ Use the potentiometer to move the cylinder.

❑ Find and record the minimum required upper (positive) voltage required to extend the cylinder.

❑ Find and record the minimum required lower (negative) voltage required to retract the cylinder.

❑ Retract the cylinder and set back the toggle switch of the proportional valve to "Ext." position.

For Example:

❑ Upper dead band of the valve = 1.09   Volt

❑ Lower dead band of the valve = -1.47   Volt

7

Second we need to apply and test the digital dead band eliminator

- ❑ Click on the "Amplifier" model and set the proportional gain to 2.

- ❑ Apply the upper and lower limits of the DB eliminator to the digital DB eliminators A and B.

- ❑ Make sure the signal is bypasses the dead band eliminator and ramp generator.

- ❑ Connect Matlab RT and run the model. Record the SSE =

- ❑ Use the potentiometer to retract the cylinder.

- ❑ Repeat the previous steps with input signal passes through DBE-A and record the SSE =

- ❑ Repeat the previous steps with input signal passes through DBE-B and record the SSE =

- ❑ Share your observation with the instructor.

8

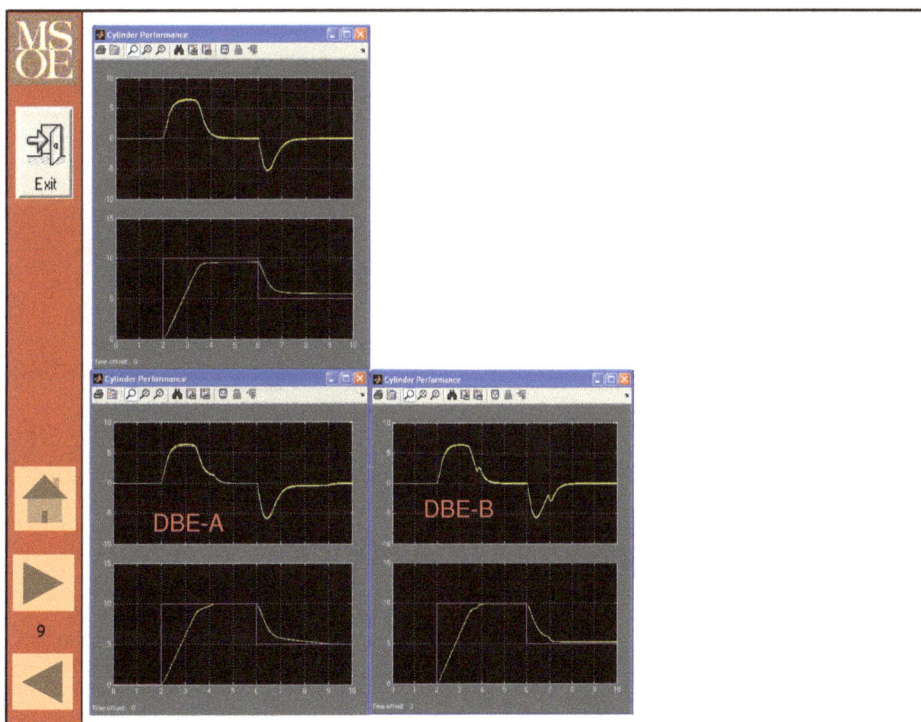

9

Step 7: Data Acquisition: (Ramp Generator)

First: observe the cylinder performance
(harshness of the start) without ramp generator

1. Click on the "Amplifier" model and set the gain to 10. Make sure the signal is bypasses the dead band eliminator and ramp generator.

2. Cylinder must be fully retracted. If not, put V1 toggle switch at "Pot." position, move the potentiometer to retract the cylinder then return back the toggle switch to "Ext." position.

3. Connect Matlab RT and run the model.

4. The model will drive the cylinder and stops after 10 second.

5. Share your observation with he instructor.

10

Second: observe the cylinder
performance with ramp generator

1. Click on the "Amplifier" model and make sure the signal go through ramp generator.

2. Cylinder must be fully retracted. If not, put V1 toggle switch at "Pot." position, move the potentiometer to retract the cylinder then return back the toggle switch to "Ext." position.

3. Connect Matlab RT and run the model.

4. The model will drive the cylinder and stops after 10 second.

5. Share your observation with he instructor.

6. Turn the Hydraulic power OFF.

7. Answer the following questions:

11

❑ Without Ramp Generator   ❑ With Ramp Generator

❑ Using the ramp generator smoothness the cylinder acceleration rate.

☐ TRUE   ☐ FALSE